Dolphins Are No

Featuring the Bottlenose Dolphin

By Stephen Schutz

Starfall Education
P.O. BOX 359, Boulder, CO 80306

Printed on recycled paper. Text: 50% post-consumer waste; cover: 10% post-consumer waste. ISBN: 978-1-59577-072-1

Did you know that dolphins are not fish?

They are different from fish in many ways.

Dolphins must come up to the surface to breathe.

Fish can breathe under the water.

Blowhole

Dolphins have one **blowhole**. They breathe in and out through it. Fish do not have blowholes.

Dolphins

Fish

Dolphins move their tails up and down when they swim.

Fish move their tails from side to side.

Dolphins

Fish

Dolphins have smooth skin.

A fish's skin is not smooth.

Dolphins are **mammals**. A mother dolphin nurses her **calf**.

Fish are not mammals.

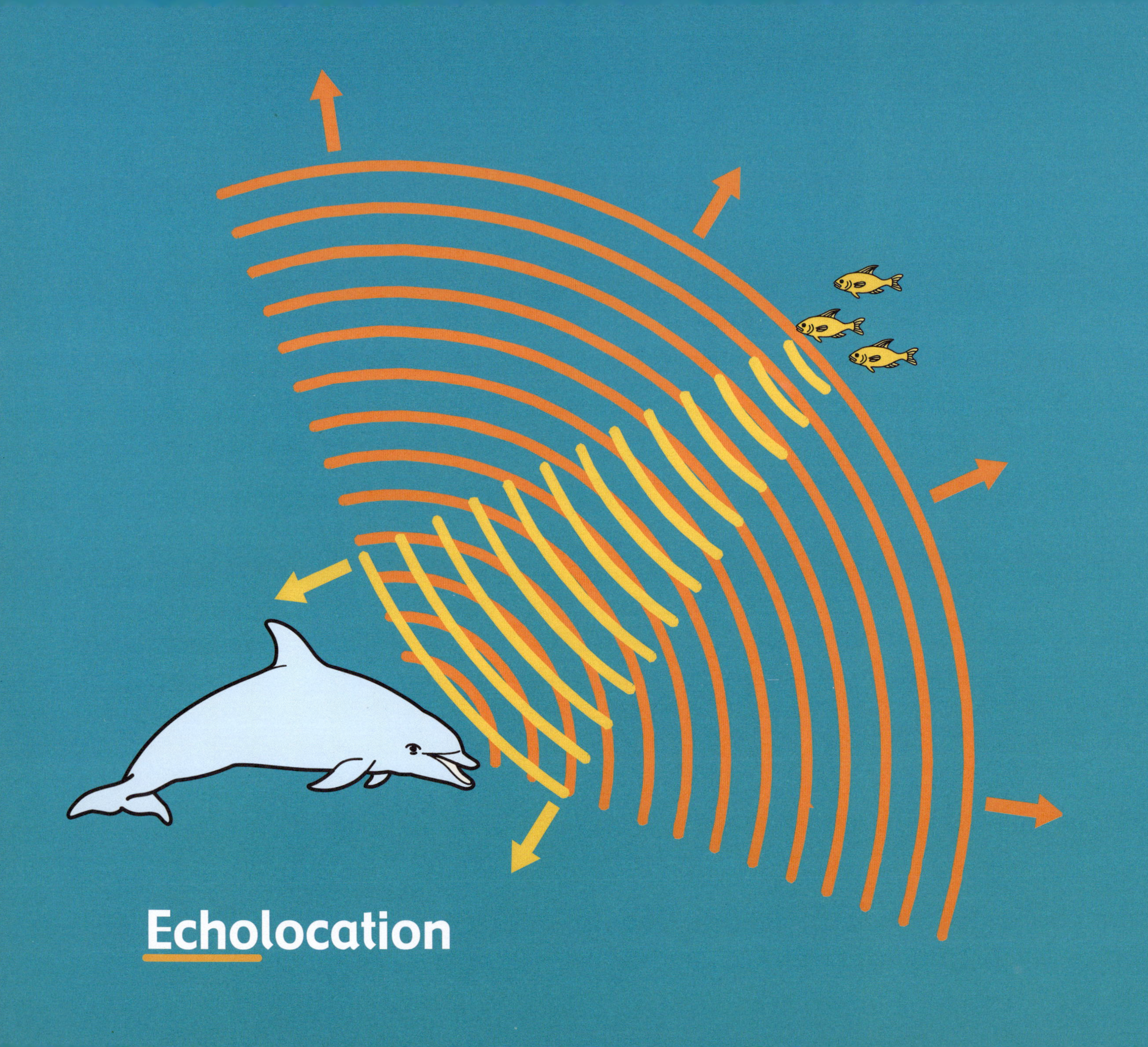
Echolocation

Dolphins "see" objects by making sounds and listening for an echo. This is called **echolocation**.

Fish cannot do this.

Dolphins are playful. They like to swim in the bow waves of ships.

Fish do not do this.

Dolphins are very smart. Some learn to do tricks!

Fish don't appear to be as smart.

Now *you* know why dolphins are not fish!

Words You Know

Blowhole

Calf

Echolocation

Mammal

Mammals feed their babies milk that comes from the mother's body. All mammals have a backbone and are "warm-blooded."

What Is a Dolphin?

If a dolphin isn't a fish, just what is it? A dolphin is a member of a group of sea mammals called toothed whales. Scientists call them toothed whales because they have teeth! In addition to teeth, all toothed whales also have one blowhole and use echolocation to find fish to eat.

The toothed whales in this book are bottlenose dolphins. Look at these other toothed whales. Even though they look different from each other they are all still toothed whales!

Pilot Whale

Orca (Killer Whale)

Beluga (White Whale)

Boto (Pink River Dolphin)

Index

About the Author

At age 9, young Stephen Schutz was still struggling to read. What came easily for some children required many more hours of Stephen's work, and he was consistently toward the bottom of his class in reading. Now a Ph.D. in physics and a successful publisher and artist, Dr. Schutz wants to make sure children in his situation today have resources that can help. He turned to the Internet and conceived Starfall.com, an online educational resource available to children the world over.

Photo Credits

The following photos were used with permission from the photographers. *Bottlenose Dolphins*, page 2, © Hiroya Minakuchi / SeaPix.com. *Dolphins swimming*, page 8, © Masa Ushioda/ SeaPix. *Wild Bottlenose Dophins mother and calf*, pages 12 & 22, © Jeff Rotman/Gallo Images/AfriPics.com. *Dolphin and ship*, page 16, © Tom Walmsley/SplashdownDirect.com.

The following images were licensed from © iStockPhoto: *Jumping Bottlenose Dolphin*, page 1, © Dirk Freder; *Goldfish*, page 10, © Viktor Kitaykin; *Playful Dolphins*, page 20, © Els van der Gun; *Pilot Whale*, page 23, © Elena Yatsenko; *Beluga Whale*, page 23, © Shirly Friedman; *Boto Dolphin*, page 23, © iStockphoto; *Girl with 2 dolphins*, back cover, © Alexander Novikov.